شجرة القَرَضْ السُوداني

الشجرة المعجزة

Sudanese Acacia Nilotica Or Qarad

The Miracle Tree

First Edition

By: Fathiya Musa Ajab

- Ph.D. of Cell Biology, Atlantic International University (AIU)

- Master of Cell Molecular biology, Long Island University (LIU)

- Bachelor of Home Science and Education, University of Khartoum (KHU)

- Adjunct Professor of biology, Touro College (NYSCAS)

- Laboratory technician of biology, Touro College (NYSCAS)

Published by : Lulu.com

ISBN 978-1-6780-4951-5

©2021 Fathiya Ajab, all rights reserved

All rights reserved. No part of this publication may be Reproduced, stored in retrieval system or transmitted in any form or by any means electronic, mechanical, photocopying, Recording or otherwise, without the prior written permission of the publisher.

In the memory of my father, my mother, my sisters, and brothers who left me in a deep sorrow, I dedicate this book as half of my Final thesis to my doctorate degree in Science majoring in Cell biology from A

CONTENTS

Abbreviations ……………………………………………………….3

Preface to the First Edition…………………………………….4

Acknowledgments……………………………………..………....5

1.0 Acacia Nilotica In Sudan north and around the world

    Summary …………………………………………………….6

    1.1 Introduction …………………………………………….8
    1.2 Acacia nilotica distribution in Africa …………….9
    1.3 Acacia nilotica distribution in Asia ……………..9
    1.4 Acacia nilotica distribution in Australia ………. 9
    1.5 Conclusion ……………………………………………..10
    References ………………………………………..………11

2.0 The Different Parts of Acacia nilotica

    Summary ……………………………………..……………12

    2.1 Introduction ……………………………………………12

    2.2 The Pods ……………………………………………….13
    2.3 The Roots ……………………………………....……..13
    2.4 The Leaves ……………………………………..……..16
    2.5 The Flowers……………………………………………16
    2.6 The Gum ……………………………………………….17
    2.7 The Seeds …………………………………………….19
    2.8 Acacia nilotica Bark……………………..………..20
    2.9 Conclusion ……………………………………………21
    References ……………………………………………….22

3.0 The Uses of Acacia nilotica

    Summary ……………………………………….…………23

    3.1 Economic Use ……………………………………….24
    3.2 Tanning Purposes ………………………………….24

    3.3 Acacia as a Medicinal Plant 27
    3.4 Conclusion... 27
    References... 29

4.0 Products of Acacia nilotica in the Market 30

Glossary.. 38

Index.. 46

Abbreviations

AIU	Atlantic International University
LIU	Long Island University
A. nilotica	Acacia nilotica
NYSCAS	New York School of Career and Applied Studies
SACNY	Sudanese American Community of New York
USA	United States of America
UK	United Kingdom
SKSn	South Kordofan, Sudan north
SN	Sudan north
P.D.R.	People's Democratic Republic of Yemen
YAR	Yemen Arab Republic
COVID 19	Coronavirus Disease 2019
HBP	High Blood Pressure

Preface to the First Edition

The editor thanks contributors who did their best to help in the editing of this book. The book is based on many researches done into this medicinal plant (*A. nilotica*). These researches are on many pathogens that causes dangerous diseases. Also, citations and interviews with some indigenous who have a long history of using these medicinal plants in herbalism. The assistance from Touro College (NYSCAS) biology department was much appreciated.

Acknowledgments

My great appreciations and Thanks to my family here in the USA (Awae Elnaw, Mazen Ajab, Yagoub Saleh) for their great help. In Sudan (Musa Sedeg, Mujtaba Ali, & Majdi Sedeg) who provided me with support, photos, and materials from Sudan north that I used in this book. Special thanks to those from the Sudanese American Community (Salwa Ahmed (SACNY member), Fakhertag Obeid (SACNY Deputy foreign office), Hind Elshazaly (The Sudanese Community Head), Salah Jumaa (SACNY Social Secretary), Adeeba Rabih (SACNY member), and Fathy Elsedeg SACNY member) for their generous support. I am very thankful for everybody's inspiration and upholding my work and enable me to succeed. My gratitude to all resources that are helpful to by providing the photos, information, or pictures of products.

1.0 Acacia Nilotica in Sudan north and around the world

Acacia nilotica is native of Sudan north where it is called Sunt or Qarad. It is famous as medicinal plant in many areas in Sudan north. People there believe that *A. nilotica* is a strong antibiotic that boost the immune system and provides protection for the respiratory and cardiovascular systems. Our parents, grandparents, and great grandparents used it for gaining good health. All along his strive life, my father never used anything else than *A. nilotica* for treating his illnesses. He used to mix *A. nilotica* pods with Karkade (Hibiscus) leaves for treating colds and influenzas. If he has any wounds, he used to apply the dry powder pods over them and gradually the wound dried up and heals in short period. Eid Aladha is the biggest Eid for Muslims that people slaughter sheep in it. The hairy skins of these animals are given as donation to traditional shoes makers, who clean them from hair by using pods powder which have high amount of tannin an astringent substance that helps in this process. The tree's wood is resistant to water specially during rainy seasons that is why it has many uses, see (3.0 in this book). Acacia nilotica in Sudan north is the most famous and common kinds of Acacia. The other two types are:

1. Acacia tortilis which known as umbrella thorn acacia or Israeli babool. Growth mostly in Africa and the middle east. This kind of acacia does not like to grow into wet areas like *A. nilotica* instead, it grows on sand mound and rocky downhills. Elephants, rhinos, Impala, and kudu love to eat it.

2. Acacia Senegalia is also called Gum Arabic tree in Sudan north. In India it is called Kher or Khor. It has many different names in other countries such as Oman, Sub-Saharan Africa, and Pakistan. The tree of this kind of Acacia is called hashab or Taleh in Sudan north and famous for their Arabic gum. Sudan north considered the first country in the world that produce and export Arabic gum. Most countries import Arabic gum from Sudan north are UK, USA, and China.

Summary

When you take the bus from the capital of Sudan north, Khartoum, traveling towards Kordofan in the west, you will be amazed by the great view of the Acacia trees spread all over the area. Below these tall trees you see the young trees covers the stems of the older trees and hide them. When you look at the higher branches of the older ones, you see the gum Arabic balls stick on them and glittering like pearls in a beautiful necklace. Their yellow flowers are tiny, clustered together in bright-yellow, round heads, surrounded by the beautiful, long constricted pods holding seeds inside them peacefully. Everybody loves to see this view and having great interest in talking about it. This area where *A*. nilotica and other types of Acacia grow, all the way along the Nile banks in the central area up to the north and grow all over the area in Kordofan and Darfur in poor and rich Savana.

Sometimes you see Arabic gum collectors collect the gum from some of these trees in a specific traditional basket called (Gofa) as we see in the picture below of a Kordofanian lady carrying Gofa and harvesting Arabic gum.

Kordofanians benefit a lot from collecting Arabic gum and selling it in the local markets cheaply and these markets export it all over the world with good prices. This process also creates hundreds of jobs for those local people. Some people believe that *A*. nilotica stops hair damage, so they use its dry leaves after mixing their powder with some water, then apply this mixture into their head, then wash it after 15-20 minutes. Other researches proved that *A*. nilotica roots have an anticancer effect.

1.1 Introduction

Acacia Nilotica (family: Leguminosae, local name in Sudan: Qarad), is a tree 5-20 meters in Hight that grows in canal banks and moist ground. The tree has long thorns as straight stipular spines in pairs. It is a well-known medicinal plant in Africa, Asia, and Australia. The different components of this plant are used in ethnomedicine in Africa and the Indian Subcontinent. A study has done by A. Rezaei in 2015 proven that A. nilotica is effective for the treatment of many diseases such as gonorrhea, cough, impotence, smallpox, ulcers, typhoid convalescence, tuberculosis, and many other diseases. *A.* nilotica acknowledged being rich in phenolics, consisting of condensed tannin and phlobatannin, gallic acid, protocatechuic acid, pyrocatechol. Several studies have proven that its different parts (roots, leaf, gum, stem bark, seeds, and pods) are useful for curing many diseases for animals and people. For example, A. Barapatre also proved that A. nilotica is safe and effective as an antioxidant and antidiabetic when used in vitro by using an extracted alkali lignin that was subjected to microbial biotransformation by ligninolytic fungus Aspergilla's flavus and Emericalla nidulas. Below is the scientific classification of *A.* nilotica (table 1)

	Scientific	Classification
Kingdom		Plantae
Clade		Angiosperm
Clade		Eudicots
Clade		Rosids
Order		Fabales
Family		Fabaceae
Clade		Mimosiodeaer
Genus		Vachellia
Species		nilotica

1.2 Acacia nilotica distribution in Africa

Acacia nilotica grows in most areas in Africa. It is grown all over African countries, especially in east African countries such as Sudan north, Ethiopia, and Somalia. In Tanzania, Ethiopia, Somalia, and Angola they depend on it completely as an important medicinal indigenous plant in Africa that grows naturally and treat them from bad demons. According to the scientific classification, it has four clades which are Angiosperm, Eudicots, Rosids, and Mimosiodeaer. The genus is called vachellia. Acacia in general has two subspecies. Acacia seyal which is widespread in tropical Africa, also has yellow flowers as nilotica but powdery white to red or yellow bark on the trunk, not dark and rough, and the pods are open. The other subspecies is Acacia karroo grows in southern Africa. It has similar flowers like nilotica but normally smooth branches and dehiscent pods. Acacia gummifera, is a rare subspecies found endemic to Morocco. It has only 1–3 pairs of pinnae and eglandular petioles to the leaves.

1.3 Acacia nilotica distribution in Asia

Acacia nilotica is widely distributed in east Asia. For example, Subsp. indica occurs in the P.D.R. Yemen, the Yemen Arab Republic, Oman, Pakistan (Punjab, Sind), India (Punjab, Uttar Pradesh, Bengal, Madhya Pradesh, Madras, Bombay) and Burma. An interview with different people from these counties showed us that this important plant has its benefit for many people around the world. It has been also cultivated in Iran, Vietnam (Ho Chi Minh) and It is spread in all these areas by livestock. In India, it is called Indica according to the name of the subspecies.

1.4 Acacia nilotica distribution in Australia

Acacia nilotica grows in north-eastern South Australia (Sydney and said to be established in Queensland). The sub of *A.* nilotica in Australia is (indica). It is thought to be native to the Indian sub-continent. There, it is known as an invasive species of significant concern as well as a weed of unspecified importance. Not much research is done there to study this M. plant but recently, Australians start to research it for more medicinal information.

1.5 Conclusion

A. nilotica proved important and efficient in many continents around the world. Its distribution in Africa, Asia, and Australia has the same characteristics, which is the areas where it grows on the canal banks. The tree is also tropical and sub-tropical tree, so it resists all different kinds of harsh weather. In Sudan north, it grows mostly in the poor savanna, rich savanna arears. The other vegetation that found in *A*. nilotica's places are short grasses in some areas and wooded long trees in another areas. To adapt to the dry areas that they live in, *A*. nilotica send its roots far away into the ground seeking sources of water and nutrients. I never forget those days when I was a child and I used to trip always by the roots of our *A*. nilotica tree in my way back home from my sister's house at night. But when years passed by, these roots have become bigger and more obvious to be seen above the ground by everybody. Now, young children use these big roots protruding above the ground as low benches to sit on in their free time for fun.

References

1- Andrew, R, Philip, J, & et all (2012). The Genus Acacia (Fabaceae) in East Africa: Distribution, diversity, and the protected area network. Plant Ecology and Evolution 145 (3): 289-301
2- Dowa8a (2019). Acacia nilotica. #Dowa8a #Journey to discover
3- https://www.cabi.org/isc/datasheetreport/2342

2.0 The Different Parts of Acacia nilotica

Summary

Interestingly, Acacia nilotica has many different parts like many other trees in rich Savana. These parts are pods, sap or gum, leaves, bark, flowers, roots, seeds, and stems. These different parts have many different uses; economically, medicinally, and spiritually. A lot of studies about *A*. nilotica are held by many researchers all over the globe. These investigations proved that this miracle plant is of a great benefit for humans, animals, and plants. It is one of the most types of Acacia used traditionally by many indigenous people, in the countries mentioned above for treating many diseases. These parts have different shapes and colors. A. nilotica trees dry during the summertime (May through August), then they start growing again by the beginning of autumn (September through December) and new pods with seeds harvested during winter (January through April). in A. nilotica usually more or less pubescent to tomentose except in the northern part of its range

2.1 Introduction

In Sudan North many people invest in selling A. nilotica for its great importance among people. There are also some manufactures for preparing it by packing it into whole pieces or powder for the general use. Acacia nilotica different parts can be used as pieces or powder and can be boiled or consumed without boiling. Most of these parts contain tannin that is why *A*. nilotica is a strong astringent. The wood has a density of about 1170 kg/m3 that is why it is used in railway tracks. The stems are very strong and resist all environmental and climate changes. They are also hard, heavy, and tough A. nilotica wood are resistant to termites that can decompose it. Its flowers are yellow fibers attached together in a circle. According to Alireza Dorostkar investigation, A. nilotica xylem is sporadic with average diameter 122 micron. It has 4.5 rays in transverse section and 12.45 xylems per mm^2. There are many products made from those different parts of A. nilotica. People use them for making many things such as sleepers, boat building, pit props, wagons, tool handles, carving, construction timbers, floor blocks.

2.2 The Pods are the most important part in *Acacia nilotica* tree

The outside of these pods is grey and light green. They are used by most people in Sudan

for many purposes. When the pods are new, they look light green and when they become old, their color become dark green to black. Structurally, there are two types of pods; one very constricted and the other has less constriction like the one in the pictures above. Each constriction has a black seed inside it. People use it as whole by (soaking it into water and add to it some pieces of Karkade or Roselle and then drink) or grind it as powder and make an aqueous solution from it or take it as powder. Many people burn the pods and inhale them to get rid of germs from all the house specially during flu time. Personally, I believe into this idea because during this COVID 19 pandemic, we always burn *A*. nilotica pods in our apartment to keep it hygienic from this virus especially when we come back from outside, and that is what happened, we are a big family from five people living in a small apartment, no body catch this disease. *A*. nilotica gargling is my favorite during this pandemic time too. It cleans the throat and chest from any congestion.

2.3 The Roots

Acacia nilotica roots are dark brown color. They are very strong and travel far into the ground. Sometimes they protrude above the ground.

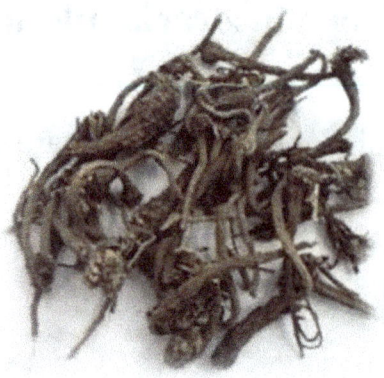

The roots are used as powder solution to treat airways infections and asthma. Also, the roots are burned like pods and inhaled traditionally for killing pathogens. In research, investigators always prefer to use small roots of young trees.

An incense burner with Qarad burning in it (Ajab Fathiya's photos).

An incense burner with Acacia nilotica pods burning for its smoke inside houses to protect from germs. (Ajab Fathiya's photos).

2.4 The Leaves

A. nilotica has green leaves arranged oppositely bi axial. They are photosynthetic that supply oxygen to the environment, the leaves are used widely for basal carcinoma by grinding them, mix the powder with water and then applied all over the infected arears. It takes only one day or two days to heal the body. The leaves also help in blood clotting and lower high blood presure (HBP)

2.5 The flowers

They are round with yellow color carried by short and thin branches. They are used by many people for treating asthma. Other people use the flowers as a drink that clear the throat.

2.6 The Gum or Sap:

Has yellow to dark brown color. It has more fibers and easy to use. Its behavior is good because it is stable in high temperature and slowly fermented. People use it traditionally to fighting cancer tumors, dysentery. It is considered an antioxidant that improve liver function and protects its inside tissues from inflammation because it contains many chemicals such as: -

a. Catechin:

Catechin is a flavan-3-ol, a type of natural phenol and antioxidant. It is a plant secondary metabolite. It belongs to the group of flavan-3-ols (or simply flavanols), part of the chemical family of flavonoids. The name of the catechin chemical family derives from catechu, which is the tannic juice or boiled extract of Mimosa catechu (Acacia catechu L.f).

b. Epicatechin:

is a compound representative of the flavanols. It can also be found in grape seeds, grape skin, tea, cola nuts, strawberries, and red wine. Some studies believe that it leads to the prevention of cardiovascular diseases in humans.

c. quercetin:

Quercetin is a plant flavanol from the flavonoid group of polyphenols. It is found in many fruits, vegetables, leaves, seeds, and grains; red onions and kale are common foods containing appreciable amounts of quercetin. Quercetin has a bitter flavor and is used as an ingredient in dietary supplements, beverages, and foods.

d. gallic acid:

Gallic acid (also known as 3,4,5-trihydroxybenzoic acid) is a trihydroxy benzoic acid with the formula $C_6H_2(OH)_3CO_2H$. It is classified as a phenolic acid. It is found in gallnuts, sumac, witch hazel, tea leaves, oak bark, and other plants. It is a white solid, although samples are typically brown owing to partial oxidation. Salts and esters of gallic acid are termed "gallates".

e. procyanidin:

Procyanidin B4 is a proanthocyanidin obtained by the condensation of (-)-epicatechin and (+)-catechin units. It has a role as an antioxidant, an EC 5.99.1.3 [DNA topoisomerase (ATP-hydrolyzing)] inhibitor and an antineoplastic agent.

2.7 The Seeds

Acacia nilotica seeds are round, hard, and black color from the outside. They contain green color flesh inside. They are used as antiviral and stop cancer tumors from spreading all over the body. They can be used soaked or powder. The seeds are found inside the pods and can be used together or separated from each other.

2.8 Acacia nilotica Bark

According to some researches. The Bark, in West Africa, used to treat cancers and/or tumors (of ear, eye, or testicles) and indurations of liver and spleen, condylomas, and excess flesh.

2.9 Conclusion

As medication started with herbal medicine, *A. nilotica* considers one of the most medicinal plants that are discovered in curing many infections a long time ago. Most parts of *A. nilotica* such as sap or gum, bark, leaves, and young pods are strongly astringent due to tannin, and are also, chewed in Sudan and Senegal as an antiscorbutic.

In late 60th, there were many widespread pandemics such as measles, smallpox, chicken pox, and tuberculosis in our hometown Dilling Sudan. What I remember is that, I and some children from our neighborhood used to play together under our *A. nilotica* tree. During the weekdays, we gather there in the evenings to play Umtagar; an interesting play by five small round stones, one flown into the air and when it lands down, you catch another stone with it. At the weekend, we play all day Aros…Aros, which means bride. We make our brides from pieces of cloths a very adorable play that we loved in our childhood. We spent most our time under that tree even during the time of those pandemic. None of us got any of those illnesses despite our contact at school with other infected children.

Nowadays a lot of researches are investigating the rich phytochemical that are found in *A. nilotica*.

References

1. *Adesokan AA, Akanji MA. Effect of administration of aqueous* extract of Enantia chlorantha on the activities of some enzymes in the small intestine of rats. Niger J Biochem Mol Biol. 2003;18(2):103–105. [Google Scholar]
2. Adzu B, Abbah J, Vongtau H, Gamaniel K. Studies on the use of Cassia singueana in malaria ethnopharmacy. J Ethnopharmacol. 2003; 88:261–267. [PubMed] [Google Scholar]
3. Agrawal S, Kulkarni GT, Sharma VN. A Comparative Study on the Antioxidant Activity of Methanol Extracts of Acacia nilotica and Berberis chitria. Adv In Nat Appl Sci. 2010;4(1):78–84. [Google Scholar]
4. Handbook on Seeds of Dry-zone Acacias FAO
5. Template: Wikispecies
6. USDA Germplasm Resources Information Network (GRIN)
7. Wickens, G.E. (1995). "Table 2.1.2 The timber properties of Acacia species and their uses". Role of Acacia species in the rural economy of dry Africa and the Near East. FAO Conservation Guide. 27. Food and Agriculture Organization of the United Nations. ISBN 978-92-5-103651-8
8. en.wikipedia.org
9. Alireza Dorostkar (2015). Investigating the Properties of "Acacia Nilotica" as a Species with Capability of Utilization in Furniture Industry. ISSN 2348 – 7968

3.0 The Uses of A. nilotica

Summary

When talking about *A.* nilotica uses, everybody remembers one of the most famous true stories happened to a man who was sick with stomach cancer in Sudan north. This man was taken to a local hospital in his town that is far from the capital Khartoum. After a lot of tests, they discovered that he has stomach cancer in a late stage, and he needs chemotherapy as a treatment. The man's stomach started to swell. So, the doctors gave him a transfer to go to Khartoum to do the chemotherapy. This patient transported from his town to the capital by a lorry filled of *A.* nilotica sacks of pods. The journey took about 12 hours to reach there. All through his trip this man was sleeping over these sacks of acacia pods. When they reached Khartoum and his family tried to get him down the lorry, he was in a perfect condition and they noticed that the color of the sacks where he was sleeping has become very red. This patient talked to his family with great joy that he is feeling better. Also, the swell in his stomach disappeared. He moved freely and easily, nobody supported him. Everybody was shocked but amazed at the same time, what has happened to him. At the hospital that he has been transferred to, the doctors told him after many diagnosing that he has nothing in his stomach. A great surprise! But after that people realize that the *A.* nilotica pods in those sacks are the secret of curing that sick man. Acacia nilotica has many benefits for humans and animals. for instance, it has economic uses, has tanning purposes, as well as medicinal plant. Each part in *A.* nilotica has its specific use. People in Africa, Asia, and Indian Subcontinent use it according to their tradition. All its parts are strong antioxidant that improves liver function. It also protects internal liver tissues from inflammation and other liver diseases. Helps in blood clotting during bleeding. Decreases high blood pressure. It also helps lowering high blood sugar. It is uninflammatory and kills human worms. It is anti-bacterial that kills bacteria that causes diarrhea and dysentery. Some people believe that *A.* nilotica stops hair damage, so they use its dry leaves after mixing their powder with some water, then apply this mixture into their head, then wash it after 15-20 minutes. Other researches proved that *A.* nilotica roots have an anticancer effect.

3.1 Economic Use

Impressively, people in SN notice that and up to date that animals such as goats, camels, and sheep eat the leaves of acacia nilotica to produce more milk. Nas stated in his study in 1980th that, the gum is good to make things such as matches, candles, paints, and inks. Also, *A.* nilotica stems are used in the railway tracks and steamships for their resistance to harsh weather and their strength. In India, the stems are used as firewood and charcoal, this species has been used in industry balers. The woods are also used as source of fuel in some industries in Sudan north. All these things considered of great advantages in addition to what is introduced at the beginning of this book from other helps to local people who cultivate *A.* nilotica gum and to the Sudanese government economy.

3.2 Tanning Purposes

Tanning with *A.* nilotica pods' powder is an old traditional process used in Sudan north for cleaning the skins of animals from their hairs or scalps of pythons and snakes. This old process is still done by experts who are called "Dabageen". In it, the skin is soaked in water first for three to four hours. Second the powder of the pods is added to the water and skin. Two or three hours later, the experts start to scrab the skin by sharp tool. This process is repeated every five hours for three days. For the python skin, they are soaked in ash solution or paint for three days first, after that they will be soaked into A. nilotica solution (for three days) to remove their scalp. These tanned leathers made ready and clean for use by manufactories. The photos below show us this process. different kinds of leathers that have tanned with A. nilotica and shoes made from them.

Tanned tiger skin for making an expensive kind of shoes called Markoob and beautiful handbags. (Ajab Fathiya's photos).

 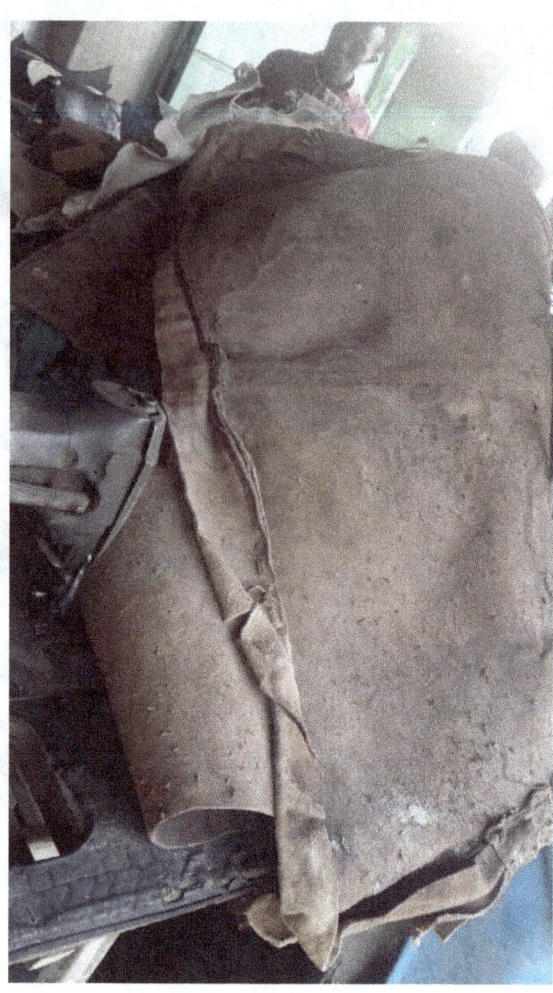

Tanned sheep skin from Dalling Sudan South Kordofan ready for sale by manufacturers. (Ajab Fathiya's photos).

Some of the products of tanned sheep skin and another picture shows an expert tanning animals' skin.

3.3 Acacia as a Medicinal Plant

It has vast traditional medicinal uses for a long time ago. It is believed that Acacia nilotica treat all kinds of common colds, dysentery, airways infection, diarrhea, gonorrheal, skin cancer, healing wounds. It is used in ethnomedicine as lactogogue. Demulcent, astringent, antibacterial and antidiabetic. A decrease in arterial blood pressure is reported by use of methanolic extract of *A*. nilotica pods in one of the advanced researches done by Atif Ali (Atif, A 2011).

3.4 Conclusion

In conclusion, the tradition uses of Acacia in many countries around the world help in improving their economy. In Sudan, for instance, Arabic gum is one of the most product that brings money to the country because it is exported to many developed countries such as USA and Europe to be used in pharmaceutical industries. Tanned leathers of Sudan are the best kinds to produce different products such as handbags, shoes, aprons, belts, wallets, purses, leather journals, passport covers, jackets and so many other products. China with its huge economy, is one of the top countries who import Sudanese leather. Furthermore, A. nilotica provides cheap and available medication for people in rural underdeveloped and developing countries. In many remote areas in Sudan where are no hospitals and pharmacies, people depend completely on *A.* nilotica for the treatment of many diseases.

References

1) Adzu B, Abbah J, Vongtau H, Gamaniel K. Studies on the use of Cassia singueana in malaria ethnopharmacy. J Ethnopharmacol. 2003; 88:261–267. [PubMed] [Google Scholar]

2) Agrawal S, Kulkarni GT, Sharma VN. A Comparative Study on the Antioxidant Activity of Methanol Extracts of Acacia nilotica and Berberis chitria. Adv In Nat Appl Sci. 2010;4(1):78–84. [Google Scholar]

3) H. Mahdi, K. Palmina and I. Glavtch (2006). Characterization of Acacia Nilotica as an Indigenous Tanning Material Of SUDAN. Forest Research Institute Malaysia.

4) Kiran Bargali and Surendra Singh Bargali (2009). Acacia nilotica: A multipurpose leguminous plant. Nature and Science, 2009;7(4), ISSN 1545-0740

https://www.jstor.org/stable/43594670

5) Atif Ali*, Naveed Akhtar, & et al (2011). Acacia nilotica: A plant of multipurpose medicinal uses. Journal of Medicinal Plants Research Vol. 6(9), pp. 1492-1496, 9 March 2012

Available online at http://www.academicjournals.org/JMPR

4.0 Products of Acacia nilotica in the Market

Currently, *A.* nilotica has many products in the market. These products found in pure form or mixed with other substances. Some are manufactured traditionally, and others are manufactured in big companies. Fig Below

Products from Asia

Great appreciation to all resources of products that I used in this book; Alamy stock photo, USDA organic, Halaleveryday.com, Alvira Natural, Herbal Treasure, Iyush herbal Ayurveda, Amalth babool, and all other resources I used in this book.

My Experience with Acacia nilotica

As one of the indigenous of S. Kordofan where A. nilotica grows all over the area, I have good relationship with it as a medicinal plant that treat many diseases. We found our ancestors benefit from it in their life. This experience inspired me to research it with many different pathogens such as gram-positive and gram- negative bacteria. I tested A.nilotica with tetrahymena thermophiles, a freshwater protozoan. A. nilotica inhibit the growth of all these pathogens. In another research I have tested it with human saliva to see its Protease Activity, antioxidant, and antibacterial effect using different procedures as you see in the Table below.

Protease	Water	Inhibitor	1	2	3	4	5	6	7	8
Positive	Clear	Negative	Negative	Negative	Negative	Negative	Negative	Negative	Negative	Negative
Negative	Clear	Positive	Negatice	Negative	Positive	Negative	Negative	Positive	Negative	Positive

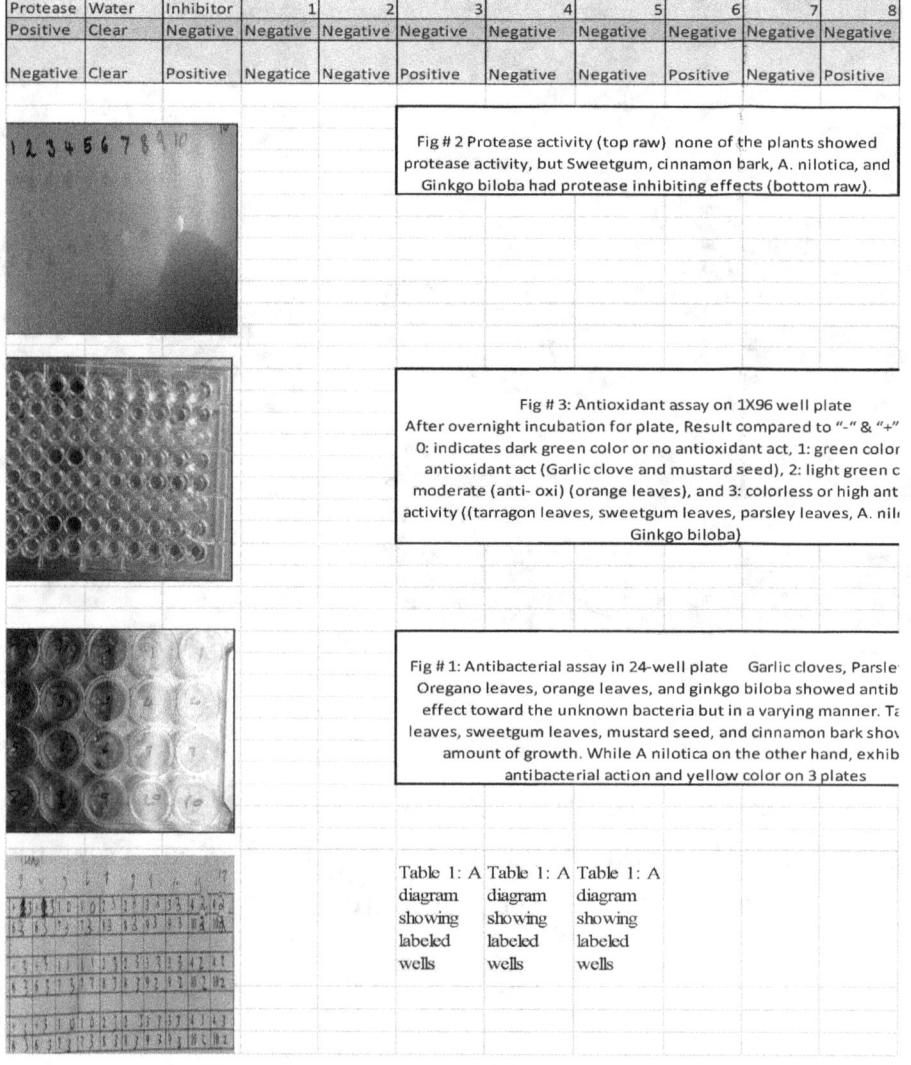

Fig # 2 Protease activity (top raw) none of the plants showed protease activity, but Sweetgum, cinnamon bark, A. nilotica, and Ginkgo biloba had protease inhibiting effects (bottom raw).

Fig # 3: Antioxidant assay on 1X96 well plate
After overnight incubation for plate, Result compared to "-" & "+" 0: indicates dark green color or no antioxidant act, 1: green color antioxidant act (Garlic clove and mustard seed), 2: light green c moderate (anti- oxi) (orange leaves), and 3: colorless or high ant activity ((tarragon leaves, sweetgum leaves, parsley leaves, A. nil Ginkgo biloba)

Fig # 1: Antibacterial assay in 24-well plate   Garlic cloves, Parsle Oregano leaves, orange leaves, and ginkgo biloba showed antib effect toward the unknown bacteria but in a varying manner. Ta leaves, sweetgum leaves, mustard seed, and cinnamon bark sho\ amount of growth. While A nilotica on the other hand, exhib antibacterial action and yellow color on 3 plates

Table 1: A diagram showing labeled wells

Table 1: A diagram showing labeled wells

Table 1: A diagram showing labeled wells

Biology: Bioassays for the Antibacterial Activity Protease and Protease Inhibition and Antioxidant Activity of Ten Selected Plants.

REPOSITORY URL:

https://touroscholar.touro.edu/researchday/2018/posters2018/13

Also, in one of my analysis for A. nilotica compared to other medicinal plants from Sudan north, against C. violencia and M. luteus Growth, A. nilotica proved very effective towards these pathogens. And finally, in my final dissertation, A. nilotica with the addition of another medicinal plant and suspending agents proved a very strong antiviral too.

Glossary

A

Acacia nilotica: a well-known medicinal plant in Africa.

Acknowledge: accept or admit the existence or truth of

Antibacterial: tending to prevent the growth or spread of bacteria.

Antidiabetic: a substance or medicine developed to stabilize and control blood glucose levels amongst people with diabetes.

Antibiotic: a medicine (such as penicillin or its derivatives) that inhibits the growth of or destroys microorganisms.

Antioxidant: a substance or compounds that inhibits oxidation. a substance such as vitamin C or E that removes potentially damaging oxidizing agents in a living organism.

Antiscorbutic: having the effect of preventing or curing scurvy.

Aprons: a protective or decorative garment worn over the front of one's clothes and tied at the back.

Assistance: the action of helping someone with a job or task.

Astringent: causing the contraction of skin cells and other body tissues.

B

Believe: accept (something) as true; feel sure of the truth of.

Benefits: something that produces good or helpful results or effects.

Boost: help or encourage (something) to increase or improve.

C

Catechin: is a flavan-3-ol, a type of natural phenol and antioxidant. It is a plant secondary metabolite.

Citation: reference to a book, paper, or author, especially in a scholarly work.

Contributors: People or things give or contribute something.

Chew: to crush, grind, or gnaw (something, such as food) with or as if with the teeth: masticate.

Citation: Machine® helps students and professionals properly credit the information that they use.

Component: a part or element of a larger whole, especially a part of a machine or vehicle.

Condense: make (something) denser or more concentrated.

Contributor: a person or thing that contributes something.

Convalescence: time spent recovering from an illness or medical treatment; recuperation.

Cure: a method or course of remedial treatment, as for disease.

D

Demulcent: a substance that relieves irritation of the mucous membranes in the mouth by forming a protective film.

Density: the degree of compactness of a substance.

Deputy: a person whose immediate superior is a senior figure within an organization and who is empowered to act as a substitute for this superior.

Diarrhea: is loose, watery stools caused by a virus or, sometimes, contaminated food.

Donation: Money or thing given as charity.

Downhill: toward the bottom of a slope.

Dysentery: infection of the intestines caused by protozoan parasite (Entamoeba histolytica) resulting in severe diarrhea with blood.

E

Editing: is the process of selecting and preparing written, photographic, visual, audible, or cinematic material used by a person or an entity to convey a message or information

Efficient: achieving maximum productivity with minimum wasted effort or expense. being productive with minimal effort.

Epicatechin: is the most abundant flavanol found in and absorbed from dark chocolate and is thought to exhibit health-promoting.

Ethnomedicine: is a study or comparison of the traditional medicine based on bioactive compounds in plants and animals and practiced by various ethnic groups, especially those with little access to western medicines, e.g., indigenous peoples.

Experts: a person who has a comprehensive and authoritative knowledge of or skill in an area.

Export: send (goods or services) to another country for sale.

F

Famous: known about by many people.

Flesh: soft tissues of an organism.

Furthermore: in addition; besides (used to introduce a fresh consideration in an argument).

G

Gallic acid is a trihydroxy benzoic acid with the formula $C_6H_2(OH)_3CO_2H$.

Glitter: particles reflect light at different angles, causing the surface.

Gonorrheal: is an infection caused by a sexually transmitted bacterium that infects both males and females.

Gratitude: the quality of being thankful; readiness to show appreciation for and to return kindness.

Gum or sap: adhesive substance of vegetable origin, mostly obtained as exudate from the bark of trees or shrubs belonging to the family Fabaceae.

H

Healing: is the process of the restoration of health from an unbalanced, diseased, damaged or unvitalized organism.

Herbal: relating to or made from herbs, especially those used in cooking and medicine.

Herbalism: is the study of pharmacognosy and the use of medicinal plants.

I

Import: to bring from a foreign or external source.

Impotence: inability in a man to achieve an erection or orgasm.

Indigenous: originating or occurring naturally in a particular place; native.

Indurations: a hardening of an area of the body as a reaction to inflammation, hyperemia, or neoplastic infiltration.

Infections: A disease caused by microorganisms that invade tissue.

Inflammation: process of fighting against things that harm it, such as infections, injuries, and toxins, to heal itself.

Inks: is the colored liquid used for writing or printing

Inspiration: the process of being mentally stimulated to do or feel something, especially to do something creative.

Interest: the feeling of a person whose attention, concern, or curiosity is particularly engaged by something.

Internal: situated near the inside of the body. Or inner parts or features.

Investigate: carry out a systematic or formal inquiry to discover and examine the facts of (an incident, allegation, etc.) to establish the truth.

J

Jackets: A piece of clothing worn on the upper body outside a shirt or blouse, often waist length to thigh length.

Journey: an act of traveling from one place to another.

K

Karkade or Roselle is a red sorrel and belongs to the family of Malvaceae.

L

Lactogogues: contain specific nutrients that support lactation.

Leather: is a durable and flexible material created by tanning animal rawhide and skins.

Lumber: also known as timber, is a type of wood that has been processed into beams and planks, a stage in the process of wood production.

M

Malaria: an intermittent and remittent fever caused by a protozoan parasite that invades the red blood cells

Material: the matter from which a thing is or can be made.

Medicinal: adj (of a substance or plant) having healing properties.

Miracle: an amazing product or achievement, or an outstanding example of something.

Moist: slightly wet; damp or humid.

N

Native: born in a particular place · belonging to a person since birth or childhood.

Necklace: an ornament worn around the neck.

O

Opposite: having a position on the other or further side of something; facing something, especially something of the same type.

P

Pathogen: any organism that can produce disease.

Pearl: Perfect shining spheres.

Pharmaceutical: relating to medicinal drugs, or their preparation, use, or sale. Or a compound manufactured for use as a medicinal drug.

phenolics: are a class of chemical compounds consisting of one or more hydroxyl groups (—OH) bonded directly to an aromatic hydrocarbon group.

phlobatannin: is - a tannin that with hot dilute acids yields a phlobaphene.

Pods: the part of Acacia plant that carry the seeds.

Protocatechuic acid: is a dihydroxybenzoic acid, a type of phenolic acid.

Pyrocatechol: is a toxic organic compound with the molecular formula $C_6H_4(OH)_2$.

Q

Qarad: is a common name for Acacia nilotica in Sudan north.

Quercitin: is a plant pigment (flavonoid)

R

Remote: situated far from the main centers of population; distant

Resists: withstand the action or effect of.

Resource: an action or strategy which may be adopted in adverse circumstances.

Rural: a geographic area that is located outside towns and cities or the countryside.

S

Soak: to become thoroughly wet by immersing it in liquid.

Species: the largest group of organisms in which any two individuals of the appropriate sexes or mating types can produce fertile offspring, typically by sexual reproduction.

Spiritual: is to be thought of as personal, and the spiritual dimension (or Holy Spirit) is universal.

Stick: to attach something to something else.

Subcontinent: a large, distinguishable part of a continent, such as North America or southern Africa.

Sudan north: is a country in Northeast Africa. It is bordered by Egypt to the north, Libya to the northwest, Chad to the west and South Sudan to the south.

T

Tannin: a yellowish or brownish bitter-tasting organic substance present in some galls, barks, and other plant tissues, consisting of derivatives of gallic acid, used in leather production and ink manufacture.

Tanning: the action or process of converting animal skin into leather by soaking it in a liquid containing tannic acid or using other chemicals.

Thorn: a stiff, sharp-pointed, straight or curved woody projection on the stem or other part of a plant.

Tracks: is the structure consisting of the rails enables trains to move by providing a dependable surface for their wheels to roll upon.

Tradition: the transmission of customs or beliefs from generation to generation.

Tumor: a swelling of a part of the body, generally without inflammation, caused by an abnormal growth of tissue, whether benign or malignant.

Tuberculosis: is a potentially serious infectious disease that mainly affects your lungs.

Typhoid: is a bacterial infection that can lead to a high fever, diarrhea, and vomiting.

U

Ulcer: an open sore on an external or internal surface of the body, caused by a break in the skin or mucous membrane that fails to heal.

Underdeveloped: not fully developed. not advanced economically.

Upholding: confirm or support (something which has been questioned).

V

Vast: of very great extent or quantity; immense.

View: a sight or prospect, typically of attractive natural scenery, that can be taken in by the eye from a particular place.

W

Wagons: a vehicle used for transporting goods or another specified purpose.

Wallets: a folding pocketbook with compartments for personal papers and usually unfolded paper money.

Weather: refers to the conditions of the atmosphere during a short period of time.

Wood: the hard-fibrous material that forms the main substance of the trunk or branches of a tree or shrub, used for fuel or timber.

Worms: any of several creeping or burrowing invertebrate animals with long, slender soft bodies and no limbs.

X

Xylem: the vascular tissue in plants that conducts water and dissolved nutrients upward from the root and helps to form the woody element in the stem.

Y

Yellow: of the color between green and orange in the spectrum, a primary subtractive color complementary to blue; colored like ripe lemons or egg yolks.

Yellowish: having a yellow tinge; slightly yellow.

Z

Zone: an encircling band or stripe of distinctive color, texture, or character.

Index

A

Acacia nilotica, 8

Adapt, 10

Agent, 19

Antibacterial, 27

Antibiotic, 38

Anticancer, 8

Antidiabetic, 38

Antineoplastic, 19

Antioxidant, 8

Antiscorbutic, 38

Antiviral, 19

Applied, 3

Appreciable, 18

Aprons, 28

Aqueous, 13

Assistance, 4

Asthma, 14

Astringent, 21

Autumn, 12

B

Basal, 16

Believe, 18

Benches, 10

Benefits, 23

Behavior, 17

Beverages, 18

Biotransformation, 8

Bitter, 18

Boost, 6

Burner, 14

C

Cancer, 17

Carcinoma, 16

Cardiovascular, 18

Catechin, 19

Catechu, 17

Chewed, 21

Citation, 39

Clotting, 25

Cluster, 7

Condylomas, 21

Congestion, 13

Constricted, 7

Contributor, 4

Convalescence, 8

Create, 7

D

Damage, 23
Dehisce, 9
Demulcent, 27
Density, 12
Deputy, 39
Dietary, 18
Donation, 6
Downhill, 6
Dysentery, 17

E

Editing, 40
Effect, 8
Efficient, 10
Eglandular, 9
Endemic, 16
Environment, 12
Epicatechin, 19
Ethnomedicine, 27
Excess, 21
Experts, 24
Export, 6
Extract, extracting, 17

F

Famous, 23
Fermented, 17
Fiber/s, 12
Fight, 17
Flavonoids, 17
Flavor, 18
Flavus, 8
Flesh, 19
Flu, 13
Fun, 10
Furthermore, 28

G

Galic acid, 28

Gargling, 13

Germs, 15

Glitter, 40

Gonorrhea, 8

Gonorrheal, 27

Gratitude, 41

Grind, 13

Gum or sap, 17

H

Hard, 19

Harsh, 24

Harvest, 7

Heal, Healing, 16

Herbal, 21

Herbalism, 4

Hygienic, 13

I

Idea, 13

Import, 28

Impotence, 8

Incense, 14

Indigenous, 36

Indurations, 21

Infections, infected, 21

Inflammation, 23

Ingredients, 28

Inhale, 13

J

Jackets, 28

Journey, 23

K

Kale, 18

Karkade or Roselle, 42

L

Lactogogues, 42

Leather, 42

Ligninolytic, 8

Liver, 17

Livestock, 9

Lumber, 42

M　　　　　　　　　**N**　　　　　　　　　**O**

Malaria, 42　　　　　　Native, 9　　　　　　Opposite, 16

Materials, 5　　　　　　　　　　　　　　　Oxygen, 16

Measles, 21

Medicinal, 23

Metabolite, 38

Microbial, 8

Mimosa, 17

Miracle, 12

Moist, 8

P **Q** **R**

Pandemic, 21

Pathogen, 36

Petioles, 9

Pharmaceutical, 28

Phenolics, 8

Photosynthetic, 16

Phytochemical, 21

Pinnae, 9

Pit props, 12

Pods, 13

Polyphenols, 18

Pressure, 23

Protruding, 10

Pubescent, 12

Purpose, 13

Qarad, 14

Quercetin, 18

Remote, 28

Resist, 10

Resource, 35

Rural, 43

S

Slaughter, 6

Smoke, 15

Soak, Soaking, 19

Solution, 24

Species, 43

Spine, 8

Spiritually, 12

Spleen, 21

Sporadic, 12

Stable, 17

Stick. 44

Stipular, 8

Strive, 6

Subcontinent, 8

Supplement, 18

Supply, 16

T

Tanned, tannin, 24

Termites, 12

Thorn, 44

Throat, 13

Timber, 42

Tiny, 7

Tomentose, 12

Tracks, 12

Tradition, traditional, 14

Trip, 23

Tropical, 9

Tuberculosis, 21

Tumor, 44

Typhoid, 44

U

Ulcer, 44

Upholding, 45

V

Vegetation, 10

View, 6

W

Wagons, 12

Wallets, 28

Weather, 10, 24

Wood, 24, 42

Worms, 23, 45

X

Xylem, 12, 45

Y

Yellowish, 44

Z

Zone, 45

Table 1

From "Medicinal Plants Use" for respiratory disorders

Type of Acacia	Family	Part Used	Traditional Use
Acacia arabica	Apiaceae	Leaves and fruits	Cough
Acacia jacquemontii	Myrsinaceae	Flower, seeds, leaves, stem, bark	Asthma
Acacia modesta Wall	Acanthaceae	Gum	Respiratory tract problems
Acacia nilotica	Apocynaceae	Flowers	Asthma
Achillea millefolium	Asteraceae	Leaves	Cold, flu

Recommended Researches on Acacia Nilotica from many Journals and Books

There are many researches about A. nilotica in different journals such as PubMed and Ethnopharmacology. Many papers have been published on the different parts of this medicinal plant. Below are some examples of these researches.

PubMed

Topics

1. Bhushette PR, Annapure US (2017). Comparative study of Acacia nilotica exudate gum and acacia gum. Int J Biol Macromol. PMID: 28390831
2. Massey S, MacNaughtan W, et al (2017). A structural study of Acacia nilotica and Acacia modesta gums. Carbohydr Polym. PMID: 28917858
3. Roozbeh N, Darvish L (2016). Acacia Nilotica: New Plant for Help in Pelvic Organ Prolapse. J Menopausal Med. PMID: 28119891
4. Al-Nour MY, Ibrahim MM, Elsaman T (2019). Ellagic Acid, Kaempferol, and Quercetin from Acacia nilotica: Promising Combined Drug with Multiple Mechanisms of Action.
PMID: 32226726 Free PMC article. Review.
5. Roozbeh N, Darvish L, Abdi F (2017). Hypoglycemic effects of Acacia nilotica in type II diabetes: a research proposal. BMC Res Notes. PMID: 28747209 Free PMC article.
6. Zabré G, Kaboré A, & et al (2017). Comparison of the in vitro anthelmintic effects of Acacia nilotica and Acacia raddiana.
Hoste H, Louvandini H. Parasite. 2017; 24:44. PMID: 29173278 Free PMC article.

7. Abdullah MAM, Farghaly MM, (2018). Effect of feeding Acacia nilotica pods to sheep on nutrient digestibility, nitrogen balance, ruminal protozoa and rumen enzymes activity.
Youssef IMI.J Anim Physiol Anim Nutr (Berl) PMID: 29363190

8. Subhaswaraj P, Sowmya M, et al (2017). Determination of antioxidant potential of Acacia nilotica leaf extract in oxidative stress response system of Saccharomyces cerevisiae.
.J Sci Food Agric. 2017 Dec;97(15):5247-5253 PMID: 28474422

9. Jadoon S, Karim S, et al (2015). Anti-Aging Potential of Phytoextract Loaded-Pharmaceutical creams for Human Skin Cell Longetivity. Oxid Med Cell Longev. 10.PMID: 26448818

Ethnopharmacology

Books

10. Saeed Abdalbasit Adam Mariod, Mohamed Mirghani and Ismail Hussein (2017). Unconventional Oilseeds and Oil Sources 1st ed. Paperback 978-0-12-809435-8. eBook 978-0-12-813433-7
11. Abdalbasit Adam Mariod (). Gum Arabic 1st ed. Paperback 978-0-12-812002-6. eBook

978-0-12-812003-3

12. Manju Singh, Deependra Singh and more (2020). Advances and Avenues in the Development of Novel Carriers for Bioactives and Biological Agents 1st ed. Paperback 978-0-12-819666-3 eBook 978-0-12-819918-3
13. Atta-ur Rahman (2020). Studies in Natural Products Chemistry, Volume 66 1st ed. Hardcover 978-0-12-817907-9. eBook 978-0-12-817908-6
14. Vidhu Aeri, D.B. Anantha Narayana and Dharya Singh (2019). Powdered Crude Drug Microscopy of Leaves and Barks1st ed. Paperback 978-0-12-818092-1 eBook

978-0-12-818225-3

15. An evidence-based Reference (2010). Natural Standard Herb & Supplement Guide 1ˢᵗ ed.

    Hardcover 978-0-323-07295-3 eBook 978-0-323-29145-3

16. Cyrus McKell (1888). The Biology and Utilization of Shrubs1st ed. eBook

    978-0-323-14361-5

17. Julian Evans, Jeffery Burley and John Youngquist (2004). Encyclopedia of Forest Sciences 1ˢᵗ ed. Hardcover ISBN: 9780121451608 eBook ISBN: 9780080548012